We Live in a Malleable Reality-and We Can Change It

Copyright Page

This book is copyrighted for 2021

We Live in a Malleable Reality-
And We Can Change It

The Time Travel and Parallel Dimensions Series Book 4

By Martin K. Ettington

All Rights Reserved USA 2021

ISBN: 9798469518815

Printed in the United States of America

We Live in a Malleable Reality-and We Can Change It

We Live in a Malleable Reality—and We Can Change It

I've had many paranormal experiences in my life and some pretty cool spiritual enlightenment experiences too. This has given me a good understanding of the spiritual development process and the development of paranormal abilities as a side effect.

I'm also an Engineer with a lot of Physics mixed into my education so I have a pretty good understanding of the conventional views of engineering and science as taught in our universities.

My fascination with the unusual and unknown also led me to write about off the wall subjects like the paranormal, time travel, and moving between dimensions. The research I did convinced me that these strange and weird phenomena do exist and are not fantasies.

All of these factors have led me to this book which has the goal of better understanding the Universe we live in, and how we can change the reality we live in through our willpower and beliefs.

My numerous experiences and research have led me to conclude that although we may think we live in a stable Universe—that is a big misunderstanding. The more I learn the more convinced I am that the Universe is much more complicated than we ever thought.

In this book I want to show you the reader that the stable reality we take for granted does not exist and we can manipulate many more things than we ever thought possible.

The more I learn, the more I realize that our Universe is a much more mysterious place than we ever thought.

We Live in a Malleable Reality-and We Can Change It

We Live in a Malleable Reality-and We Can Change It

Other books by Martin K. Ettington

Spiritual and Metaphysics Books:
Prophecy: A History and How to Guide
God Like Powers and Abilities
Enlightenment for Newbies
Removing Illusions to Find True Happiness
Using the Scientific Method to Study the Paranormal
A Compendium of Metaphysics and How to Guides (Six books together in one volume)
Love from the Heart
The Enlightenment Experience
Learn Your Soul's Purpose
Pursuing Enlightenment
A Modern Man's Search for Truth
Use Intuition and Prophecy to Improve Your Life
The Handbook of Spiritual and Energy Healing

Longevity & Immortality:
Physical Immortality: A History and How to Guide
The Commentaries of Living Immortals
Records of Extremely Long Lived Persons
Enlightenment and Immortality
Longevity Improvements from Science
The 10 Principles of Personal Longevity
Telomeres & Longevity
The Diets and Lifestyles of the World's Oldest Peoples
The Longevity Six Books Bundle
Long Lived Plants and Animals

Science Fiction:
Out of This Universe
The Immortals of the Interstellar Colony
The Mystic Soldier

The Immortality Sci Fi Bundle
Visiting Many Universes
The History of Science Fiction and Fantasy

The God Like Powers Series:
Human Invisibility
Invulnerability and Shielding
Teleportation
Psychokinesis
Our Energy Body, Auras, and Thoughtforms
The God Like Powers Series— Volume 1 Compilation

The Yoga Discovery Series:
Yoga-An Ancient Art Form
Hatha Yoga-Helping you Live Better
Raja Yoga-Through the Ages
The Yoga Discovery Package

Business & Coaching Books:
Creating, Paublishing, & Marketing Practitioner Ebooks
Building a Successful Longevity Coaching Business
Why Become a Coach?
The Professional Coaching Success Trilogy
2020-Make Money Writing and Selling Books
The 2020 Handbook of High Paying Work Without a College Degree
The important of Creativity and How to Improve Yours

We Live in a Malleable Reality-and We Can Change It

Science, Technology, and Misc.
Future Predictions By and Engineer & Seer
The Unusual Science & Technology Bundle
The Real Atlantis-In the Eye of the Sahara
Removing Limits On Our Consciousness-And Thinking Outside the Box

Survival
Survival of Humanity Throughout the Ages
33 Incredible True Survival Stories
The Importance of Fire in History and Mythology
How to Survive Anything: From the Wilderness to Man Made Disasters
Building and Stocking a Nuclear Shelter for less than $10,000
The Human Survival Five Books Bundle

Legendary Animals and Creatures
Are Cryptozoological Animals Real or Imaginary?
Fire in History and Mythology
All About Dragons
Sea Serpents and Ocean Monsters
The Legendary Animals Five Books Bundle

Ancient History
The Real Atlantis-In the Eye of the Sahara
Ancient & Prehistoric Civilizations
Ancient & Prehistoric Civilizations- Book Two
The History of Antediluvian Giants
The Antediluvian History of Earth
Ancient Underground Cities and Tunnels
Strange Objects Which Should Not Exist
More Out of Place Artifacts
Strange and Ancient Places in the USA
A Theory of Ancient Prehistory And Giant Aliens
The Destruction of Civilization About 10,500 B.C.

Aliens and Space
Aliens and Secret Technology
Aliens Are Already Among Us
Designing and Building Space Colonies
Humanity and the Universe
All About Moon Bases
All About Mars Journeys and Settlement
The Space and Aliens Six Books Bundle
A Theory of Ancient Prehistory and Giant Aliens
The Space Colonies and Space Structures Coloring Book
All About Asteroids
Spaceships, Past, Present, and Future
Astronauts, Cosmonauts, and Other Important Space Flyers
All About Mars Journeys and Settlement
Mining the Asteroid Belt

Time Travel and Dimensions
Real Time Travel Stories From a Psychic Engineer
The Real Nature of Time: An Analysis of Physics, Prophecy, and Time Travel Experiences
Stories of Parallel Dimensions
We Live in a Malleable Reality-and We Can Change It

We Live in a Malleable Reality-and We Can Change It

<u>The Longevity Training Series</u>

(A transcription of the online Multimedia Longevity Coaching Training Program)

The Personal Longevity Training Series-Book1-Long Lived Persons
The Personal Longevity Training Series-Book2-Your Soul's Purpose
The Personal Longevity Training Series-Book3-Enable Your Life Urge
The Personal Longevity Training Series-Book4-Your Spiritual Connection
The Personal Longevity Training Series-Book5-Having Love in Your Heart
The Personal Longevity Training Series-Book6-Energy Body Health
The Personal Longevity Training Series-Book7-The Science of Longevity
The Personal Longevity Training Series-Book8-Physical Body Health
The Personal Longevity Training Series-Book9-Avoiding Accidents
The Personal Longevity Training Series-Book10-Implementing These Principles

The Personal Longevity Training Series-Books One Thru Ten

These books are all available in digital and printed formats from my website and on Amazon, Barnes & Noble, Apple ITunes, and many other sites

My Books Website is: <u>http://mkettingtonbooks.com</u>

We Live in a Malleable Reality-and We Can Change It

Signup for our Mailing List to get the following:

1) A discount coupon for 25% discount on all books on our site
2) Occasional Notices of new books available
3) Occasional Email on other offerings of ours (Monthly)

Go to this link to sign-up:
http://personal-longevity.com/mkebooks/emailsignup/

And click this link to get the FREE 102 page Ebook titled "Secrets of Many Things"

If you have any questions about this book or other subjects please contact the Author at:

mke@mkettingtonbooks.com

We Live in a Malleable Reality-and We Can Change It

Table of Contents

1.0 Introduction 1
2.0 The Limits of Our Knowledge 3
Part One-Hidden Aspects of Reality 7
3.0 Premonitions and Prophecy 7
   3.1 My Intuitions 7
   3.2 Additional Experiences of Intuition and Prophecy 17
4.0 Time Travel Interfaces 21
   4.1 Visiting Versailles in the Past 21
   4.2 Bold Street, Liverpool, England 25
   4.3 Sir Victor Goddard 35
5.0 Stories of Multiple Dimensions 39
   5.1 A Vanishing Road 39
   5.2 Swap with Self 41
   5.3 The Man from Taured 43
6.0 Quantum Mechanics 45
7.0 Holes in Reality 49
8.0 The Underlying Spirit of the Universe 51
Part Two-Change Yourself and the World 55
9.0 Thoughtforms and Visualizations 55
   9.1 Thoughtforms 55
   9.2 My Radiation Shelter 59
10.0 Changing Our Futures 63
   10.1   Learning to Meditate or Pray Deeply 63
   10.2   Visualizing the Future 64

We Live in a Malleable Reality-and We Can Change It

- 10.3 Checking the Future for Problems 67
- 11.0 Surviving Impossible Situations 69
  - 11.1 Prisoners Escaped and Walked 4,000 Miles 69
  - 11.2 Odette Sansom 1942-1945 in Captivity 71
  - 11.3 Ernest Shakelton's Voyage to the South Pole 77
- 12.0 Spiritual and Energy Healing 85
- 13.0 Spiritual Development and Psychic Abilities 89
  - 13.1 The Yoga Sutras of Patanjali 89
  - 13.2 How Superhuman Abilities work 93
  - 13.3 The Path to Supernormal Abilities 95
- 14.0 Summary 99
- 15.0 Bibliography 101

We Live in a Malleable Reality-and We Can Change It

1.0 Introduction

I've had many paranormal experiences in my life and some pretty cool spiritual enlightenment experiences too. This has given me a good understanding of the spiritual development process and the development of paranormal abilities as a side effect.

I'm also an Engineer with a lot of Physics mixed into my education so I have a pretty good understanding of the conventional views of engineering and science as taught in our universities.

My fascination with the unusual and unknown also led me to write about off the wall subjects like the paranormal, time travel, and moving between dimensions. The research I did convinced me that these strange and weird phenomena do exist and are not fantasies.

All of these factors have led me to this book which has the goal of better understanding the Universe we live in, and how we can change the reality we live in through our willpower and beliefs.

My numerous experiences and research have led me to conclude that although we may think we live in a stable Universe—that is a big misunderstanding. The more I learn the more convinced I am that the Universe is much more complicated than we ever thought.

In this book I want to show you the reader that the stable reality we take for granted does not exist and we can manipulate many more things than we ever thought possible.

We Live in a Malleable Reality-and We Can Change It

The more I learn, the more I realize that our Universe is a much more mysterious place than we ever thought.

We Live in a Malleable Reality-and We Can Change It

2.0 The Limits of Our Knowledge

One of the things we need to realize is that we will never learn everything about the Universe we live in. Here is more information as to why:

Our Limited Understanding of Reality

Our scientific understanding of the world is only about 500 years old. Mankind has existed for over 100,000 years, and the universe is billions of years old.

Is humanity so arrogant as to say that we have a close to final understanding of the natural scientific laws of the universe, or should we be more humble and admit that we only understand a tiny fraction of what is out there, and much more is undiscovered than discovered.

I spent many years reading articles and journals from organizations like the ASPR (American Society for Psychical Research) who have done good experimental work for 60 years on validating and understanding psychic phenomenon.

However if I were to go to the average person on the street they would say that these things have never been proven.

(I also read the standard scientific journals like Science and Scientific American.) Most scientists would also say that paranormal events haven't been proven to exist.

Instead of exploring how to understand and benefit from these abilities, most researchers in these areas are still being asked to prove that these things really exist. In this case many of the skeptics aren't really interested in the objective evidence because it would disrupt their cozy worlds.

We Live in a Malleable Reality-and We Can Change It

The Scope of Reality

From the foregoing discussion on the scientific method and what is measurable, you can tell that I must have a significantly different idea of reality than the norm. The Diagram above best illustrates my belief of our ability to understand Reality:

The inner yellow circle represents what we can measure with our instruments today and perform experiments on to prove or disprove theories.

The red circle is a larger area, which we will eventually be able to measure to understand and prove or disprove the way things are

We Live in a Malleable Reality-and We Can Change It

The blue outer circle is the largest area, and is that part of the universe which we may be able to experience but will never be able to measure and validate with objective scientific approaches.

We may be able to subjectively perceive a lot of things in the blue area, but will never have the tools and techniques to objectively quantify it.

This blue realm may also include such things as where the soul goes after death, the fundamental nature of God, and certain dimensions of space and time, which we can postulate but never prove or disprove.

The red region may be more amenable to creative approaches for objective measurement and validation.

However, there will have to be agreement among the scientific community on some new approaches, which may constitute legitimate standards for objective measurement of phenomena.

This may include indirect evidence, which is used in areas like particle physics.

Neutrinos for example can't be directly perceived, but their existence can be inferred by collisions with other particles, which make cloud tracks which we can directly perceive.

The same approach should be transferable to validation of something like telepathy, where the medium of thought transference may not be understood at this point, but it can be validated through well-controlled blind studies and statistics.

We Live in a Malleable Reality-and We Can Change It

I think that this type of validation issue of the objectivity of an experiment also presents barrier to further scientific progress.

Until new objectivity standards are set, we will never make good progress on a scientific understanding of consciousness and "non-physical" phenomenon.

We Live in a Malleable Reality-and We Can Change It

Part One-Hidden Aspects of Reality

3.0 Premonitions and Prophecy

In this chapter I'm sharing some of my major personal experiences. These experiences provide a basis for my subjective analysis of how prophecy works.

3.1 My Intuitions

These examples are also intended to show the reader that I do know what I'm talking about when it comes to having "sensed" the future.

a. Visions

During the summer of 1975 I had a summer CO-OP job at General Electric's Gas Turbine engineering group in Schenectady, NY

At this time I used to meditate at my desk during the lunch hour.

We Live in a Malleable Reality-and We Can Change It

One day in early August I was meditating and thinking about a trip I was planning to Cape Cod. My mind was wandering as I was thinking about what I would do there. My thoughts went to what I would do at the beach.

All of a sudden, I had a blinding flash of a scene where I was in the surf at the beach, and a surfboard was coming towards me. Then a shock occurred and I was thrown out of my meditation and was wide-awake.

I thought that this was pretty weird, and mentioned this to a friend or two.

Two weeks later I was walking on the beach on Cape Cod. I saw a couple of guys with surfboards and asked where I could rent one to give it a try.

They said they had an extra one and I could try it with them. (I had totally forgotten my meditation vision at this point)

I tried to get up on that board all day, and had some modest success, but I was also getting exhausted in the process.

I decided to try it again and fell off when a big wave hit me. Next thing I knew I was coming up to the surface and I saw the exact same scene from my meditation.

The board hit me hard in the chin and almost knocked me out. I staggered to the shore and the two guys I was with helped me to the hospital where they put 10 stitches and 2 sutures into my chin.

We Live in a Malleable Reality-and We Can Change It

The question arises—Would I have been able to avoid the accident if I had remembered my vision and not gone surfing?

Later experiences have convinced me that the future is a set of probabilities, and we have free will to decide our actions.

I also had an experience on that trip of being able to partially heal my wounds very quickly through a deep meditation and application of psychic healing techniques. However, I do still have a small scar on my chin from this accident.

b. Warnings of Danger

<u>Detroit</u>

In 1980 I was moving from Dekalb, Illinois to Rochester, NY between assignments at General Electric, Inc.

While staying with my cousin outside Detroit, I made arrangements one evening to meet an old RPI friend Steve. We decided to go into downtown Detroit to the newly completed Renaissance Center to eat dinner and look around.

The Renaissance Center was built near the water and surrounded by slums.

After dinner we were walking out through the lower level in an area that was all boarded up with nobody else there.

Suddenly, I had this strong urge to turn around and go to find a restroom. I stopped walking forward because the urge was so strong.

We Live in a Malleable Reality—and We Can Change It

I tried to walk forward again and again a very strong urge came to turn around and go back into the main center where other people were. I remarked to Steve that I couldn't go forward—that something wouldn't let me.

Just then two black guys in trench coats appeared about 30 feet away from behind one of the foundation pillars we were about to walk past. They started walking towards us with smiles pasted on their faces.

My friend Steve took off running back into the main area and after a moment or two I figured I didn't know what these guys were carrying under their coats, so I ran too.

In less than 30 seconds we were back in a populated area with Police present, and the two guys chasing us gave us smiles like "next time we'll get you" and took off going the other way.

I had previously always tried to pray to God for protection, and tried to give a subconscious message to my senses to warn me of danger.

I'm convinced that whatever sense or "angel" warned me that evening, I would have been killed or severely wounded if I had continued walking out of the complex with no warning.

We Live in a Malleable Reality-and We Can Change It

At the Border in El Paso

In December of 1987 my Dad drove out to Houston to accompany me on my move to Los Angeles where we were going to start a business together.

The first night on the road we stayed overnight in El Paso, Texas near the border.

Sunday morning I suggested we stretch our legs by taking a walk down to the Border which appeared to be less than a mile.

As we were walking through a rundown area near the border I had a strong sense that we were in danger. This sense continued for several minutes that somebody wanted to hurt us.

I told my Dad we needed to turn around and he agreed.

We got back to the hotel safely, and even though we didn't see any danger, I'm convinced my extra senses picked up something.

We Live in a Malleable Reality-and We Can Change It

Planning a Trip to Spain

During early August of 1998, my wife and I decided to send her and our kids to visit her mother in Barcelona, Spain.

I was going to buy a ticket separately, and meet them there during early September.

When I started to call the travel agent to book my ticket I had a terrible feeling of fear about taking the flight.

I tried two other times to book the ticket during the week for a September 2nd departure, and each time I got the same strong feelings of fear and death.

I have always prayed and tried to guard myself mentally to avoid disasters, so finally I took the warning seriously and decided not to go at all.

We Live in a Malleable Reality-and We Can Change It

This was very difficult to do since I really wanted to see my wife and kids, and this meant I would be home alone for a month.

Work wasn't an excuse either, since I wasn't doing any really heavy contract work at the time and could easily have taken the time off.

I called my wife and told her my decision, and she was surprised, but agreed for me to follow my instincts.

On September 2nd the Swissair disaster occurred on a plane leaving Kennedy airport in New York, which crashed in Newfoundland Canada with all lives lost.

I would not have originally been booked on that flight, but could have easily ended up on it since I was due to fly through Kennedy airport, and any delay might have caused me to switch planes.

I will never know for sure, but this was a very strong warning.

I should also mention that for several years before this event I had strong feelings that my I would be killed in the near future. After this happened those feelings ended.

We Live in a Malleable Reality-and We Can Change It

A few Seconds Ahead

Sometimes just having sensitivity about what will happen a few seconds into the future will have a positive effect.

Avoiding a car accident at an intersection is one result.

I believe animals have spiritual abilities too.

Here is an example concerning our last dog Apollo. Some years ago he was watching my wife as she planned to start disconnecting a motor inside our dishwasher.

Apollo started barking madly at us (which he never did) and we suddenly realized that we hadn't turned off the power to the dishwasher.

He seemed to be sensing a future event which made him really worried.

We Live in a Malleable Reality-and We Can Change It

c. <u>Dreams of Indian Ocean Tsunami in 2004</u>

Back around the year 2000 through 2004 I was having a series of dreams which were similar but all slightly different.

I seemed to be in a tropical coastal area and at some type of resort. There were lots of people on the beach and there were different types of resorts in each dream.

I had a lot of fear and then it happened. There would be some type of huge wave which crashed over us, or the tide would go out and a huge wave would come in and I would be covered by the wave.

When that happened I never escaped but seemed to be one of the victims.

I recall this type of dream happening at least five to ten times over that multi year period.

Of course the disaster finally happened—the late December 2004 Tsunami of the Indian Ocean which killed at least one quarter million persons.

We Live in a Malleable Reality-and We Can Change It

Scientists researching this event now think this may have been the worst Tsunami disaster in over 600 years in the Indian Ocean.

I have since had a few dreams about Tsunamis hitting Southern California where I live, but none recently…

We Live in a Malleable Reality-and We Can Change It

3.2 Additional Stories of Intuition and Prophecy

a. Experiences about 9/11/2001

There are many reported experiences of premonitions about the major disaster of terrorists striking the two towers on 9/11/2001. Here are a couple of my own experiences: In 1976 during the summer, I had spent the last couple of years going through a spiritual and psychic development class on the side while going to college. My parents lived in New Jersey and we used to go into Manhattan once in a while. This was while the Port Authority towers were still under construction.

I was very interested in them and decided to go to the site to see how far up I could go in the towers. At the time half the building was finished and the public could go up in the elevators. I was able to get to about the fiftieth floor. I entered the floor and it was empty except for the main columns. Going over the window I wondered how long these towers would last? I decided to try my intuition to get an answer. So I put my hands on one of the pillars next to the windows, closed my eyes and asked "How long will this

We Live in a Malleable Reality-and We Can Change It

building last?" I figured it would be at least one hundred years. Imagine my shock when my intuition said to me "Twenty Five years". This didn't make any sense at the time so I figured I must be wrong.

In retrospect this experience showed me that our prophecy abilities can tell us about events many years in the future.

Then in the year 2000 in September I was in New York City again visiting my sister who lived there. I was married at the time and my ex-wife had some intuitional abilities. We were pushing a stroller with our baby son in it along with my sister who had her baby daughter in her stroller. The location was in the financial district next to the twin towers. Suddenly my ex-wife said "I just got a strong impression that many people are going to die here." This is all the detail I remember. We soon forgot the incident and went on to enjoy our visit with my sister.

Some events are so large that they affect the past and many people even in different times around them.

We Live in a Malleable Reality-and We Can Change It

A Strong Intuitional Experience

I used to live on a Peninsula south of Los Angeles called the Palos Verdes Peninsula. At the time I lived in an apartment near the cliffs. Several times a week I liked to take a walk on the path along the cliffs and around a resort next door. It was about a two mile walk. One morning I was walking down the pathway on the grass towards the cliff path. I was in a good mood and just letting my consciousness drift while I enjoyed the outdoors.

All of a sudden I started getting feelings that I should jump over the cliff. What I thought? Why am I getting these impressions? I am not a suicidal person. The thoughts continued as I headed for the cliff path and I had to struggle a little to have my regular walk. Then I saw it- hundreds of yards down the cliff path were a bunch of people holding a service. They were congregated at a location where I knew there was a memorial with flowers for a women who had taken her life by jumping only a few weeks before.

We Live in a Malleable Reality-and We Can Change It

It was a big "Ah" moment for me as I realized I was picking up the thoughts of these persons who were all thinking about why this woman had jumped. As I realized this the thoughts of jumping in my own mind disappeared. This showed me that if our minds are really open we can pick up impressions from others and may often do so-even if we don't realize it.

We Live in a Malleable Reality-and We Can Change It

4.0 Time Travel Interfaces

The following real time travel stories are from my book "Real Time Travel Stories from a Psychic Engineer". These are stories I researched which show that the reality we perceive probably has many more holes and time warps in it than we think.

4.1 Visiting Versailles in the Past

Here is a famous story of two women in 1901 who had a visit to Versailles which shifted to the year 1792

Moberly and Jourdain recounted that they had decided to visit the Palace of Versailles as part of several trips around Paris, detailing how, on 10 August 1901, they travelled by train to Versailles. They remembered not thinking much of the palace after touring it, so they said they decided to walk through the gardens to the *Petit Trianon* but after reaching the *Grand Trianon* found it was closed to the public.

We Live in a Malleable Reality-and We Can Change It

They recollected traveling with a Baedeker guidebook, but said they became lost after missing the turn for the main avenue, *Allée des Deux Trianons* and entered a lane, where they bypassed their destination. Moberly reported that she noticed a woman shaking a white cloth out of a window while Jourdain recalled noticing an old deserted farmhouse, outside of which was an old plough.

At this point they described a feeling of oppression and dreariness coming over them after which men who they thought looked like palace gardeners told them to go straight on. Moberly described the men as "very dignified officials, dressed in long greyish green coats with small three-cornered hats." Jourdain recalled that she noticed a cottage with a woman holding out a jug to a girl in the doorway, describing it as a "*tableau vivant*", a living picture, much like Madame Tussauds waxworks.

Moberly did not observe the cottage, but remembered that she felt the atmosphere change. She wrote: "Everything suddenly looked unnatural, therefore unpleasant; even the trees seemed to become flat and lifeless, like wood worked in tapestry. There were no effects of light and shade, and no wind stirred the trees."

The Comte de Vaudreuil was later suggested as a candidate for the man with the marked face allegedly seen by Moberly and Jourdain.

They reported reaching the edge of a wood, close to the *Temple de l'Amour*, and coming across a man seated beside a garden kiosk, wearing a cloak and large shady hat. According to Moberly, his appearance was "most repulsive... its expression odious. His complexion was dark and rough." Jourdain noted "The man slowly turned his

We Live in a Malleable Reality-and We Can Change It

face, which was marked by smallpox; his complexion was very dark. The expression was evil and yet unseeing, and though I did not feel that he was looking particularly at us, I felt a repugnance to going past him. They said that another man whom they described as "tall... with large dark eyes, and crisp curling black hair under a large sombrero hat" came up to them, and showed them the way to the *Petit Trianon*.

Moberly said she noticed a lady sketching on the grass who looked at them after they crossed a bridge to reach the gardens in front of the palace. She later described the lady as wearing a light summer dress and a shady white hat with lots of fair hair. Moberly reported that she thought she was a tourist at first, but the dress appeared to be old-fashioned. Moberly came to believe that the lady was Marie Antoinette. Jourdain, however, did not see the lady.

At their return to the palace, they reported that they were directed round to the entrance and joined a party of other visitors. They said that after they toured the house, they had tea at the *Hotel des Reservoirs* before returning to Jourdain's apartment.

We Live in a Malleable Reality-and We Can Change It

We Live in a Malleable Reality-and We Can Change It

4.2 Bold Street, Liverpool, England

Bold Street is a site with the most experiences of time slips so far discovered. The below stories indicate that there is some type of time warp on this street which some people pass through.

The Liverpool Time Slips and Mysterious Occurrences in Bold Street are numerous. This location seems to be some type of time portal. Several stories follow.

The Bold Street Timeslips Liverpool parascience.org.uk

The subject of time has always intrigued us. Is it as set as we have always believed? Or does time loop back on itself, giving us a glimpse of a shadowy past out of the corner of our eye.

We Live in a Malleable Reality-and We Can Change It

Was is just our imagination that made us believe we had seen an object or building change before our very eyes, and seem as though we had stepped back into the past? When this happens we usually shake our heads and put it down to imagination.

But over the last few decades, something strange has been happening in or near Bold Street, Liverpool England. Not just a glimpse of the past, but full immersion into the strange and mysterious world of English History, if only for a few moments at a time.

The strange thing about the Bold street time slips is the actual time and place they are set. In the following cases, the people involved do not go back really far, but seem to visit a particular decade or decades.

So far, most of the sightings have centered around the 1950s and '60s. This is strange in itself. Most time travel experiences seem to take the recipient back to the 18th or 19th century. But not in this case.

Are these people simply copying each other in their experiences, or are they genuinely taking a step back in time?

The answer to this has to take into account whether they are doing it deliberately to get noticed. In other words are the perpetuating a hoax?

Another explanation could be mass hallucination.
And last but not least, they really are experiencing this strange phenomena!

We Live in a Malleable Reality-and We Can Change It

The most important point is, the very first person that had this experience, obviously totally believed in what he saw, heard and felt.

So, does time flow like a river? Or does it twist and turn, going forward then sweeping back, picking up historic events and placing them down in front of you, if only for a few moments?

In this first tale, we find Frank and his wife out for a stroll in Liverpool town center. It is 1996.

His wife decided that she wanted to go and buy a book at Waterstone's the large book store, and they started to walk towards the area of the shop.

As they approached Bold Street, Frank decided to go to another shop first, but bumped into his friend, and stopped to chat in the street. His wife went ahead without him.

We Live in a Malleable Reality-and We Can Change It

A few moments later, Frank said goodbye, visited his shop and turned to go back to meet his wife. After reaching Bold Street, he headed on towards the bookstore. As he approached, he glanced up and was surprised to see the name, Cripps above the door. As he was about to cross over to see what was going on, a van swept past him with the name Cardin's on the side. The van driver honked his old fashioned horn and drove past.

Looking around, Frank suddenly realized that things were not quite what they should be. He looked at the cars driving past and realized that they were all old fashioned vehicles such as people would drive back in the 50's and 60's.

And then he noticed the people. Men were wearing hats and macs, and the women were dressed in head scarves, full skirts and had old fashioned hair styles such as women wore just after the war.

By this time, Frank was beginning to feel slightly freaked out. He carried on crossing the road and headed towards the store.

As he got closer he noticed in the window there were handbags, shoes, and umbrellas. Suddenly he saw a young woman looking up at the shop sign. She looked confused.

She was wearing modern clothes and as she saw him approaching, she smiled at him.

We Live in a Malleable Reality-and We Can Change It

Frank went into the shop, closely followed by the young woman. When they entered he was surprised and pleased to see that it had indeed turned back into a bookshop. The young woman smiled, shook her head and said, 'that was strange, I thought it was a new clothes shop!' then she walked away looking extremely puzzled.

This may sound an unlikely tale, but the odd thing about it is that Frank was, in fact, a former Police officer who was used to dealing in facts, and definitely wasn't the type of person who would believe in the paranormal.

Frank never stopped talking about it. Was this a time slip? Evidently Cripps was a women's shop that sold clothes and other goods decades before!

And Cardin's was also a well-known Liverpool firm that owned vans around the same time.

We Live in a Malleable Reality-and We Can Change It

The second story concerns a young girl by the name of Imogen. She had decided to go into Liverpool to buy her sister Abigail a few things for her new baby. Upon arriving she was happy to see a new MotherCare store that had opened up on the corner of Lord Street and Whitechapel.

She wandered around the store, and picked up a few baby items such as cardigans, baby bibs, and gloves. She was surprised to see how cheap the items were, but thought they were on offer as the store had just opened. Taking them to the counter, she tried to pay with her credit card. The staff member looked at her suspiciously, and went off to get the manager.

When she came back, she looked at the card and told Imogen that they didn't take cards. So, disappointed, Imogen went and put the items back as she hadn't any money with her.

When she got home, she told her mother what had happened. Her mother was surprised and really puzzled. 'That store closed years ago,' she said. 'There is a bank there now, in fact that's where I have my account'. Not believing her, Imogen took her mother back to the same place the next day. Sure enough the store wasn't there. It was a bank, just as her mother had told her.

The third tale is of a young man named Sean, who, while shoplifting in Liverpool back in 2006, ran away from a Security Guard and headed down Hanover Street. Trying to shake off the Guard, Sean, 19, turned into a dead end street called Brookes Alley.

By this time he was out of breath and started to get a tight sensation in his chest. He soon realized that actually it

We Live in a Malleable Reality-and We Can Change It

wasn't a problem with him, but the atmosphere around him.

He waited for the Guard to come around the corner after him, but he never appeared. So, thinking he had given him the slip, he sauntered back out and started to walk down Hanover Street again. But he soon realized that something was wrong.

The road looked different, and so did the pavement. He noticed cars driving by that looked very old fashioned, and the road works that he knew were there, were now gone. Soon he saw that the people around him were wearing strange clothes. Crossing over to Bold Street, he noticed that there were traffic lights where they weren't before, and bushes growing around the Lyceum, near a bar that he recognized.

He carried on walking. Soon he began to feel that something was not quite right. Then he began to panic. He realized that somehow he had stepped back in time. And the time slip was not going away.

Then he remember his Cell phone. Taking it out of his pocket, he tried to get a signal, but of course it didn't work. Eventually he began to really panic, but soon spotted a kiosk selling newspapers and headed over.

Leaning over the Stand, he took a look at the front page of the Daily Post. There in bold lettering was the date. 18th May 1967.

He wondered what to do. What happens if he can't get back to his own time? What about family and friends? So, speeding up his pace, he reached H. Samuel the Jewelers, and tried his phone once again. This time it

We Live in a Malleable Reality-and We Can Change It

worked. Sighing with relief he looked around and realized that he had returned to the present. But the strange thing was, he could still see, down the end of the road, people still walking around in 1967.

By this time Sean had seen enough, and dived onto a bus to go home. When he was interviewed by the local newspaper later, he stated over four times, the exact account.

Now, you may think that Sean was making the story up to escape from the guard. But the strange tale didn't end there. When the Security Guard was interviewed, he stated that when he ran after Sean, and turned down the dead end alley after him, he said that Sean had completely disappeared!

When the newspaper checked out the facts of Sean's story, they found that everything he said was historically accurate.

We Live in a Malleable Reality-and We Can Change It

These three stories are just the tip of the iceberg. There are many tales from around Liverpool that tell of time slips, ghosts and other strange phenomenon. The stories keep coming thick and fast, and of course the more tales, the more likely people will start to believe that they are all being made up, or as the saying goes, Urban Tales. So, what do you think? Real life time slips, imagination, mass hallucination or purely tales that have started out as fun but have turned into the greatest Urban Legends of all time.

My opinion is that, yes, something did happen.

Probably to the first guy, Frank who was just out shopping with his wife. The others? Maybe it was a case of mistaken roads, taking a wrong turning or just a glitch in the person's memory. By the time they get home they totally believe what happened.

Or is it true? There are so many cases concerning Bold Street, and just about anywhere else in Liverpool, that maybe, just maybe they are all living on top of the biggest time slip phenomena in the World.

We Live in a Malleable Reality-and We Can Change It

We Live in a Malleable Reality-and We Can Change It

4.3 Sir Victor Goddard

A Flight through Time

Sir Victor Goddard's trip into the unexplained involved an airplane flight. This was a much more personally harrowing experience.

In 1935, while a Wing Commander, Goddard flew a Hawker Hart biplane to Edinburgh, Scotland, from his home base in Andover, England, for a weekend visit. On the Sunday before flying back, Goddard visited an abandoned airfield in Drem, near Edinburgh, this location being closer to his final destination than the airport at which he landed. The Drem airfield, constructed during the First World War, was a shambles. The tarmac and four hangars were in disrepair, barbed wire divided the field into

We Live in a Malleable Reality-and We Can Change It

numerous pastures, and cattle grazed everywhere. It was now a farm, and completely useless as an airfield.
On Monday, Goddard began the flight back to his home base. The weather was dark and ominous, with low clouds and heavy rain. Goddard was flying in an open cockpit over mountainous terrain without radio navigational aides or cloud flying instruments. Rain began beating down on his forehead and onto his flying goggles badly which obscured his vision. He thought he could climb above the clouds, but he was wrong. He made it to 8,000 feet, looking for a break in the clouds. There was none.

Suddenly Goddard lost control of his plane. It began to spiral downward. He struggled with the controls. He could speed up or slow down, but he could not stop the spin. He was unsure of his location, but knew he was falling rapidly and might smash into the mountains before coming out of the clouds. The sky became darker, the clouds turning a strange yellowish-brown. The rain came down even more heavily. Goddard's altimeter showed he was only a thousand feet above the ground and dropping rapidly. At two hundred feet and still spiraling downward, he began to see a bit of daylight through the murky gloom, but his spiral toward seemingly inevitable death was far from over. Goddard was now flying at 150 miles per hour. He emerged from the clouds over "rotating water" that he recognized as the Firth of Forth. He was still falling.

Suddenly, he saw directly before him a stone sea wall with a path, a road, and railings on top of it. The road seemed to be slowly rotating from left to right. The cloud cover was down to forty feet. Goddard was now flying below twenty feet and was within an instant of tragedy. A young girl with a baby carriage ran through the pouring rain. She ducked her head just in time to avoid Hart's wingtip. Goddard succeeded in leveling out his plane after that. He barely

We Live in a Malleable Reality-and We Can Change It

missed striking the water after clearing the sea wall by a few feet.

He was now flying only several feet above a stony beach. Fog and rain obscured all distant visibility, but Goddard somehow located his position. He identified the road to Edinburgh and soon was able to discern, through the gloom, the black silhouettes of the Drem Airfield hangars ahead of him, the same airfield he had visited the day before. The rain became a deluge, the sky grew even darker, and Goddard's plane was shaken violently by the turbulent weather as it sped toward the Drem hangars-and into a different world.

Suddenly, the sky turned bright with golden sunlight. The rain and the farm had vanished. The hangars and the tarmac appeared to have somehow been rebuilt in a brand-new condition. There were four planes lined at the end of the tarmac. Three were standard Avro 504N trainer biplanes; the fourth was a monoplane of an unknown type- the RAF had no monoplanes in 1935. All four airplanes were bright yellow. No RAF airplanes were painted yellow in 1935. The airplane mechanics were wearing blue overalls. RAF mechanics never wore anything but brown overalls when working in hangars in 1935.

It took Goddard only an instant to fly over the airfield. He was only a few feet above the ground-just high enough to clear the hangars-but apparently none of the mechanics saw him or even heard his plane. As he sped away from the airfield, he was again engulfed by the storm. He forced his plane upward, flying at 17,000 feet and then, for a time, at 21,000 feet. He managed to return to his home base safely.

We Live in a Malleable Reality-and We Can Change It

Goddard felt elated when he landed. He then made the mistake of telling fellow officers about his eerie experience. They looked at him as if he were drunk or crazy. Goddard decided to keep silent about what had happened to him. He did not want a discharge from the RAF on mental grounds.

In 1939, Goddard watched as RAF trainers began to be painted yellow and the mechanics switched to blue coveralls. The RAF introduced a new training monoplane exactly like the one he had seen in his flight over Drem. It was called the Magister. He learned that the airfield at Drem had been refurbished.

Another twenty-seven years went by, but Goddard never forgot what had happened. He played it through over and over in his mind. It was not until 1966 that he wrote of this experience. Over the years he had become convinced that there was no way he could have known that the RAF would change the colors of their trainers and their mechanics' overalls four years before these changes took place. Goddard finally concluded that he must have glimpsed the future-or even traveled into it-for a brief moment in time.

We Live in a Malleable Reality-and We Can Change It

5.0 Stories of Multiple Dimensions

My Book "Stories of Parallel Dimensions" has many stories I found from my research which indicate real events happening between ours and multiple dimensions. These stories are distinctly different than time travel stories.

5.1 A Vanishing Road

(Here is a story told in the first person) About 4 years ago, I lived in this fairly small flyspeck of a town. At the time, I had lived there for about 12 years so I knew my way around. Our house was about a mile and a half away from the nearest neighborhood.

Our mom intentionally picked that house due to the lack of neighbors. It was tucked away on a back road, with the woods surrounding it. Every now and again, I liked to take walks with my little brother, who at the time was about 13.

We decided to do just that. We headed up the road and decided to try to find a new path or a new clearing that we

We Live in a Malleable Reality-and We Can Change It

hadn't discovered yet, when we noticed something a little shocking. Just off the road that led almost directly to the neighborhood, there was a brand new paved road. Every road in that part of town was a gravel road, so seeing an out of place paved road was pretty unusual. We stared at it for a while, and came to the conclusion that it must have been made within the last few days, due to the modern but slow growth of the town. However, we had no explanation for how they did it so fast.

We decided to explore it a bit. I remember as soon as we set foot on the road, the air became notably colder, by at least 5 degrees. The road itself was a black pavement, but no dividing lines. It was surrounded by some thick, red trees that resembled redwoods, but they were too short and non-native to our state (southern Arkansas). We walked on the road for about 3 miles until we decided to head back due to it getting dark. When we got off the road, we felt the temperature go back up. My brother and I agreed to explore it the next day.

At roughly noon the following day, we set back out to explore this place, only to discover that the entire road was now missing. When I say missing, I mean the trees that were cleared to make it had apparently grown back, with no sign of the redwood-like trees. We even began to explore the woods once more, but only to find no sign that it ever existed. When we asked our parents about it, they said they knew nothing about any new road work being done near us.

We Live in a Malleable Reality-and We Can Change It

5.2 Swap with Self

Eerie but also really cool so posted by the reddit user "ajknox09". He shared the time he swapped places with his other self so first it's important to know that his wife to be passed away when he was 20. down the line he got married to another woman who he's been with for 20 years and they have two daughters together so one day they were all at a restaurant when all of a sudden he became very dizzy even grabbed the table to brace himself when he turned to look at his wife he saw that it wasn't his current wife.

No in her place was his 20 year old wife to be and their daughter who looked just like her then he was overcome with a dizzy sensation again and when it stopped he was back with his current wife which means that in another reality his wife-to-be never passed away and they went on to have a life together which also means he's technically married to two different women at the same time like is that considered cheating if your other self has a different partner?

We Live in a Malleable Reality-and We Can Change It

We Live in a Malleable Reality-and We Can Change It

5.3 The Man from Taured

On a seemingly normal day in 1954, a seemingly normal man allegedly flew into Tokyo, but upon landing at the Tokyo International Airport, his seemingly normal trip had taken a very drastic turn for the weird. When he handed over his passport to be stamped, the man was immediately interrogated as to the whereabouts of his origins. It wasn't a case of racial profiling: While his passport looked authentic, it listed a country no one had ever heard of called Taured.

The mystery man claimed his country was located between France and Spain, but when he was asked to point it out on a map, he pointed to the Principality of Andorra. Insisting he had never heard of Andorra and that Taured had existed for 1,000 years, he claimed that he was in Japan on business, something he had been doing for the past five years. His passport seemed to back up his story, as it was covered in previous customs and visa stamps,

We Live in a Malleable Reality-and We Can Change It

and he carried with him legal currency from several European countries. He even had a driver's license issued by the mysterious country and a checkbook containing checks from an unknown bank.

After more interrogation and confusion for both parties, the traveler was sent to a nearby hotel until an official decision could be reached. There, two immigration officials stood outside the hotel door until morning. It was then that they discovered the mystery man had vanished without a trace, which was troubling, since the only possible exit was a window with no ledge 15 stories above a busy street. The Tokyo police department conducted an extensive search but continually came up empty-handed. Hopefully, if he really was from a parallel Earth, he was able to find a way back to the comforts of his home in Taured.

We Live in a Malleable Reality-and We Can Change It

6.0 Quantum Mechanics

There is reason to think that our thoughts may affect reality. Here is the story of a quantum experiment which may show that this is true:

An odd space experiment has confirmed that, as quantum mechanics says, reality is what you choose it to be. Physicists have long known that a quantum of light, or photon, will behave like a particle or a wave depending on how they measure it. Now, by bouncing photons off satellites, a team has confirmed that an observer can make that decision even after a photon has made its way almost completely through the experiment—seemingly well past the point at which it would become either a wave or a particle. Such delayed-choice experiments might someday probe the fuzzy frontier between quantum theory and relativity, researchers say.

Other researchers have demonstrated the same counterintuitive effect in the laboratory. But the new work shows that a photon's nature remains undefined even over thousands of kilometers, says Philippe Grangier, a

We Live in a Malleable Reality-and We Can Change It

physicist at the Institute of Optics in Palaiseau, France, who collaborated on an earlier test. "It's a very nice experiment that demonstrates their ability to do quantum physics in space."

A photon can act like a bullet like particle or rippling wave—but not both at once—depending on how experimenters decide to measure it. In the late 1970s, famed theoretician John Archibald Wheeler realized that experimenters could even delay the choice until the photon had made its way almost completely through an apparatus configured to emphasize one property or the other, thus proving that the photon's behavior isn't predetermined.

Wheeler imagined sending photons one at a time through a so-called Mach-Zehnder interferometer, which accentuates light's wave nature. Using a mirrorlike "beam splitter," the interferometer splits the entering photon's quantum wave in half and sends the two waves along different paths, like people walking opposite ways around the block. A second beam splitter then recombines the waves, which interfere with each other to shunt the photon toward either one of a pair of detectors. Which detector is triggered depends on the difference in the two paths' lengths, as expected for interfering waves.

Remove the second beam splitter and interference becomes impossible. Instead, the first beam splitter sends the photon down one path or the other, like a particle. As the paths cross where the second beam splitter would have been, the detectors click with equal probabilities regardless of the paths' lengths. Wheeler realized that experimenters could even wait to remove the second beam splitter until after the photon had passed the first beam splitter. That assertion suggests, weirdly, that a decision in

We Live in a Malleable Reality-and We Can Change It

the present determines an event in the past: whether the photon split like a wave or took one path like a particle.

Quantum theory avoids the issue by assuming that, until measured, the photon remains *both* a particle and a wave. Now, a team led by Francesco Vedovato and Paolo Villoresi of the University of Padua in Italy has performed a version of the experiment using the 1.5-meter telescope at the Matera Laser Ranging Observatory in southern Italy to bounce photons off satellites thousands of kilometers away. At such distances, physicists cannot make light take two parallel paths, Villoresi notes, as the spreading beams would overlap and merge. Instead, they send a photon through a Mach-Zehnder interferometer on Earth that has paths of very different lengths. The difference in path lengths splits the single pulse into two, separated in *time* by 3.5 nanoseconds, which the telescope then shoots skyward.

Once the pulses return, the experimenters run them back through the interferometer. The apparatus can either undo the time shift so that the two pulses overlap and interfere like waves or double it so that no interference is possible.

Of course, the physicists must choose which thing happens. When the pulses first leave the interferometer, they have different polarizations. To undo the time shift, physicists must first use a very fast electronic polarization to change their polarization in a certain way. To double the time shift, they simply leave their polarizations alone.

When experimenters make the pulses overlap, the photon triggers one detector or another with a probability that depends on the satellite's recession speed, as expected for interfering waves. When the pulses cannot interfere, then the photon, like a particle, ends up in either detector

We Live in a Malleable Reality-and We Can Change It

with a 50-50 probability regardless of the satellite's speed. Crucially, physicists choose which measurement to make after the light pings off the satellite halfway through its 10-millisecond round-trip, they report 25 October in *Science Advances*. Again, the delayed decision seems to reach back in time, defining how the photon behaved after it left the first beam splitter.

The experiment isn't the most stringent test of Wheeler's idea, notes Jean-François Roch, a physicist at the École Normale Supérieure in Paris, who in 2007 led a more faithful test. For example, to see the light at all over such long distances, Villoresi and colleagues must fire pulses containing many photons, instead of the individual photons Wheeler specified. Still, Roch says, the experiment is a noteworthy example of taking "quantum optics" out of the lab and into space. In May, physicists in China used a satellite to establish a weird quantum connection called entanglement between two photons sent to widely separated cities.

Delayed-choice experiments could help probe the boundary between relativity—which requires that cause precede effect—and quantum theory, Roch says. Even though, strictly speaking, the effect does not violate causality, it still raises a tension by suggesting that a measurement in the present shapes what can be inferred about the past. "This area where you mix quantum mechanics and relativity is still relatively unexplored," Roch says, "and this is the sort of experiment that raised the possibility of probing the link" between the two.

We Live in a Malleable Reality-and We Can Change It

7.0 Holes in Reality

My experiences show there are definitely "holes" or unexplainable events in our reality. Here is one hard to explain event I experienced:

I woke up one morning in July 2020 and went to the living room about 5AM. There was an Iphone on the table which was unusual since it wasn't my Iphone and my son has a different brand. It also had a rubber cover. My first thought was that my son's girlfriend had left it there during the middle of the night because it looked similar to hers.

Picking it up I saw there were messages on the screen and some type of picture behind the obscuring the messages. Not wanting to spy I put it back in the same place on the table.

My son got up a little later, so I asked him "did your girlfriend come in the middle of the night to surprise you?" He looked at me strangely and denied her visiting at night. When I asked him if he was sure he started to get upset and said "YES".

Later at 9AM the cleaning lady came. I looked and the phone was no longer on the table. I asked her where was

We Live in a Malleable Reality-and We Can Change It

the phone which had been on the table? She said it was hers and she just put it in a different location. I asked her if I could look at it and she said "YES". So I picked up that phone. It was the same one which had been on the table at 5AM. I looked at the screen and it was the same too. So how did the cleaning lady's phone get to my table at 5AM when she didn't arrive until four hours later? I know this was not a normal event or can be explained away by conventional explanations.

The Iphone was real. I picked it up, held it, and its display worked just like a real Iphone.

I've experienced many paranormal experiences in my life, but nothing like this one. This was really out of the box, even for me.

We Live in a Malleable Reality-and We Can Change It

8.0 The Underlying Spirit of the Universe

My experiences and learning have led me to believe that our consciousness is part of and linked to the underlying substrate of the Universe. Here are more of my thoughts and information on this view of reality:

The Reality of Stillness

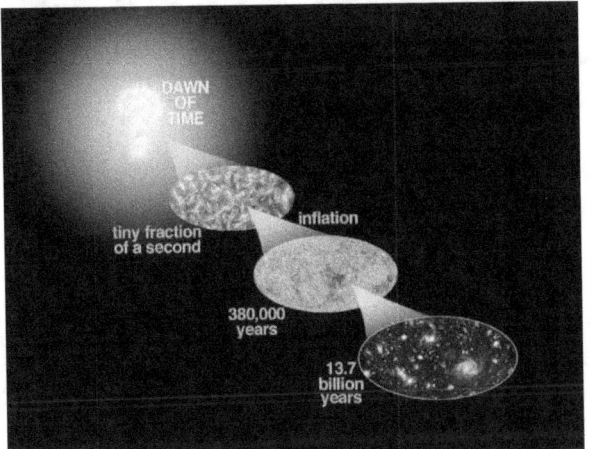

The Growth of the Universe

Most people believe that God was the initial creative force which started the Universe.

Physicists and Astronomers all agree that the Universe we know was created from nothing and inflated in a huge explosion called the "Big Bang". As it inflated time and space as we know them came into existence.

When you study Einstein's Relativistic physics you being to understand that time and space are inextricably linked. You can't have one without the other.

We Live in a Malleable Reality-and We Can Change It

Given our understanding of physics, we know that time and space didn't exist before the Big Bang. The state of things before creation then was "No Time & No Space".

A NOTIONAL PICTURE OF A BLACK HOLE

Another subject of great interest to astrophysicists is what are called "Black Holes". Black Holes are a result of Einstein's equations and astronomers have verified their existence in the last few decades.

Black Holes are stars which due to their own mass have collapsed down to an infinitely small point and where time stops. Scientists do not understand where all that mass goes.

Hmm.... A Black Hole seems to be another example of part of reality that exists without time and space.

In Quantum Physics, time is also viewed differently than we perceive it on a daily basis. Here is a quote from a Physics website explaining this view:

We Live in a Malleable Reality-and We Can Change It

> The upshot is that, on the microscopic level, there is no direction to time -- and this is even more spectacularly true in quantum physics than in classical physics. In the microscopic domain, everything just exists in a kind of nebulous, atemporal continuum. Then, every once in a while, something becomes observable, and enters the one-dimensional time continuum. The arrow of time does not exist in the universe as a whole. It only exists in individual subjective views of the universe!

I think it is fair to say that the place of stillness where time and space don't exist is part of our reality.

Therefore, it shouldn't be considered too strange that our immortal spirit is part of and one with that stillness.

We Live in a Malleable Reality-and We Can Change It

We Live in a Malleable Reality-and We Can Change It

Part Two-Change Yourself and the World

In Part 2 of this book I'm covering a lot of practices that individuals can do to modify the Universe for themselves.

9.0 Thoughtforms and Visualizations

We can create energy focuses which are called thoughtforms, and our visualization processes can also affect reality. Here are explanations of the thoughtform and an example of a visualization process I used to build a shelter:

9.1 Thoughtforms

A **thoughtform** is a manifestation of mental energy, also known as a *tulpa* in Tibetan mysticism. Its concept is related to the Western philosophy and practice of magic.

A number of *prima facie* unrelated definitions have been suggested:

An image or images held in the mind of a practitioner which aids in the manifestation of intent.

An agency of psychic effect which exists and takes form on the pre-physical realms of existence, which acts in accord with the Intent of its creator(s).

A living spiritual being created by humans. It could be a magical person's helper, or a being created by the belief in it from masses of people.

A homunculus of awareness: an Instantaneous observer / observed duality. They are created by everyone every moment (in some formulations they are everyone every moment); and they possess wills of their own.

We Live in a Malleable Reality-and We Can Change It

As we become more sensitive by spiritual development and opening of our third eye, we can learn to see thoughtforms too as part of the aura details.

The book "Thought-forms" By C.W. Leadbeater and Annie Besant is a classic study done over one hundred years ago and I highly recommend it for an in depth treatment of this subject. Here are some pictures from that book that their research said represented certain types of thoughts or emotions:

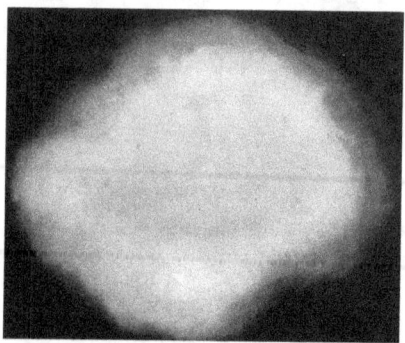

The above thoughtform represents intellectual pleasure

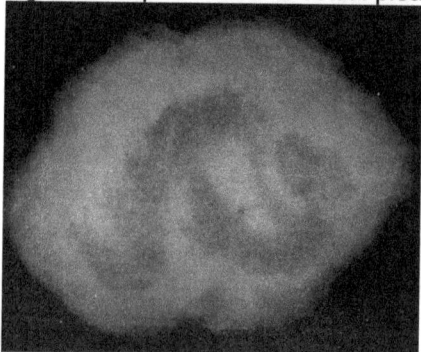

This above image indicates vague selfish affection

We Live in a Malleable Reality-and We Can Change It

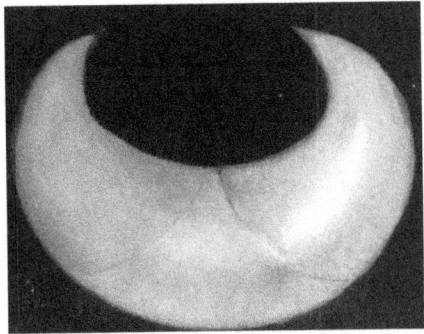

The more complex image above represents love, peace, and protection.

As you learn more about thoughtforms and how to interpret them you will be able to understand much more complex images

We Live in a Malleable Reality-and We Can Change It

We Live in a Malleable Reality-and We Can Change It

9.2 My Radiation Shelter

Here is a chapter from one of my survival books titled "Building and Stocking a Nuclear Shelter for Under $10,000" which demonstrates the importance of the visualization process to create and build something:

Designing and Visualizing the Shelter

I'm one of those people who believe in visualizing a goal before I start a project. It helps psychologically to put your mind in order, and I think it also helps moving energies in the direction of the goal.

So I spent the first several weeks after I decided to build the shelter visualizing it being built, and what it would look like.

I did this until one day the shelter just seemed solid in my mind, I knew that it would be built and stopped this intense visualization process.

The Physical Design

The design I came up with looked like this diagram:

We Live in a Malleable Reality-and We Can Change It

The design had an entrance hallway, water tanks in the back of the hallway, and then the main room to the left of the hallway.

My home was on a hillside and there is a nice corner in the back left side of my yard where a shelter building could be constructed. The shelter was also to be built well into the hillside. The only parts which would really be exposed would be the front wall, the entrance doorway, and the roof. The rest would be covered.

I visualized digging out a pit in the hillside, laying concrete, putting in the rebar, building the cinderblock walls, and then finishing the structure. I imagined what the whole structure would look like inside and outside when it was finished.

Radiation Calculations

Having been trained as a Nuclear Engineer I knew a fair amount about how to mitigate radioactive fallout. And this was my biggest concern because my property did not directly face the harbor so it would be protected from a

We Live in a Malleable Reality-and We Can Change It

nuclear blast, but would definitely get a lot of radioactive fallout in the air from winds over a nuclear blast area.

What I wanted to know was how to calculate the thickness for the roof and front wall to stop incoming radiation. These would also be made of different materials which absorb radiation at different rates. These materials included cinderblock, concrete, dirt, gravel and railroad ties. (Wooden ties in the front wall)

Searching the internet, this led me to a spreadsheet which would calculate radiation transfer through different materials and the thickness needed. (See the bibliography) The goal of my calculations was to determine the thicknesses of materials which would be needed to protect the people inside the shelter for a couple of months until the estimated radioactive fallout outside would go away.

The shelter was designed to support a family of four with supplies and water for two months. It was assumed that a person would occasionally need to go outside to dispose of trash and waste and would need to change clothes and wipe off after being in the fallout outside.

From this information I calculated about four feet of combined materials would need to be on top of the roof, and in the front wall of the shelter.

The steps to do this calculation are the following:

1) What is the likely size of a nuclear detonation and how far from your location?

2) Determine how far from a nuclear detonation radioactive dust in the air intensity will be and how long it should last. There is federal information available on

We Live in a Malleable Reality-and We Can Change It

fallout from a bomb blast of a certain size and the REMS which will be in the atmosphere until the half-life deteriorates to a safe level. In my case it was two months until the fallout from a reasonably sized bomb would dissipate in the air.

3) What is the radiation absorption for different materials. This tells you what thicknesses you will need in these materials to cover your shelter.

We Live in a Malleable Reality-and We Can Change It

10.0 Changing Our Futures

The future is all probabilities as demonstrated by my story from chapter three about how I changed my future to avoid a plane flight which crashed. Here are some things we can learn to affect these probabilities:

10.1 Learning to Meditate or Pray Deeply

First, it is key to be able to get in touch with your spirit, and the best way to do that is through learning meditation or deep prayer techniques.

Meditation is the key to developing most spiritual abilities since it allows one to calm down the mind and start to perceive your higher self or spirit. It is this spirit which we are aiming to connect with.

Meditation is best learned through an instructor. These days there are many instructors to choose from in the major cities.

We Live in a Malleable Reality-and We Can Change It

We Live in a Malleable Reality-and We Can Change It

10.2 Visualizing the Future

Once you are in a very relaxed state your consciousness will not be distracted.

The next thing to do is to start visualizing yourself in a future scene as accurately as possible.

When I had the vision described earlier in this book about the surfboard accident I was meditating deeply. While meditating I was going through my mind about places I intended to visit on the vacation trip I was planning later in the summer. While thinking about the beach on Cape Cod and what I might do there like surfing was when I had the vision.

I must have tuned into a major probable event in my life because then I was just there and saw the scene exactly as it later happened—from the viewpoint of being in my own body and the accident happening to me.

We Live in a Malleable Reality-and We Can Change It

We Live in a Malleable Reality-and We Can Change It

10.3   Checking the Future for Problems

One of the habits I've acquired in my life is to look ahead before I take a trip somewhere. I do it so often it has become a habit.

When I had the warnings of danger about buying plane tickets to Spain, I was thinking about the upcoming trip, the flight over there, and what I would do there.

The Danger warning I received came as kind of a "dark cloud of death" that gave me a lot of fear whenever I started thinking about buying the plane tickets. I have never had a reaction like that before or since when planning to buy travel tickets.

Another story illustrates what happens when you don't follow your intuition or haven't defined it in detail.

We were planning a family trip to Disneyworld for the spring of 2008. Prior to the trip I did my usual review of what we would be doing. For some reason I had the impression of a black border around Epcot--not a deep dark fatal feeling. Just the feeling that something would go wrong that would be a personal problem.

I should have delved deeper into the things we might do there. Maybe I would have gotten a more specific warning- and saved myself a very uncomfortable day.

When we were there I agreed to take the "Orange Ticket" ride on the Space Ride and became very motion sick. Then I threw up on the ride and had to go to the infirmary for several hours.

We Live in a Malleable Reality-and We Can Change It

This was a situation that could have been avoided if I'd had just a little more forethought and review of the planned activities there.

We Live in a Malleable Reality-and We Can Change It

11.0 Surviving Impossible Situations

Surviving impossible situations is all about believing and acting like you really know you can survive. Here are some stories of people caught in deadly situations and how they survived from my book "33 Incredible Survival Stories".

11.1 Prisoners Escaped and Walked 4,000 Miles

"The Long Walk" is a popular term that is attached to one of the most famous prison escape attempts of the 20th century. After being imprisoned in the far north of Russian Siberian wasteland, Polish soldier **Slawomir Rawicz** and six of his friends managed to escape from the gulag and go on incredible journey in which they walked over **4,000 miles** southward until they reached safety in India. Events from this miraculous journey were described in Rawicz's popular autobiographic book, but many modern historians dispute some portions of his tale.

According to the Rawicz's book "The Long Walk", he was imprisoned by the soviets as a political prisoner after the events of the German-Soviet invasion of Poland. In 1941 he reached distant Siberian gulag, where he and five

We Live in a Malleable Reality-and We Can Change It

friends managed to escape from the prison during a strong blizzard and start their journey south. The group consisted of Rawicz, two Polish soldiers, Latvian landowner, a Lithuanian architect, American engineer and during the journey they were joined by the 16 year old young Polish girl. During the journey they traveled south avoiding Russian towns, and eventually walked through Gobi Desert, Tibet, and the Himalayas, until they have finally reached British India in March of 1942, some 11 months after their escape. Four of the group members died on a journey - two in the Gobi and two in the Himalayas.

Some historical records that were found after Slawomir Rawicz's book was released in public contradict his tale. According to the Soviet records, Rawicz was released from the Gulag camp in 1942, before going directly to Iran. But, there are also witness reports of group presence in India. British Intelligence Captain Rupert Mayne told the media that during his post in India he debriefed three men in who claimed to have escaped from a Siberian Gulag camp. Although many historians disputed the accuracy of the book "The Long Walk", it slowly became a world bestseller and was translated in over 25 languages. In 2010 American director Peter Weir released the movie "The Way Back" (starring Jim Sturgess, Colin Farrell and Ed Harris) that adapted the tale of Slawomir Rawicz on his **journey to freedom.**

We Live in a Malleable Reality-and We Can Change It

11.2 Odette Sansom 1942-1945 in Captivity

Sansom made a landing on a beach near Cassis on the night of 2 November 1942, and made contact with Captain Peter Churchill, who headed Spindle, an SOE network based in Cannes. Her code name was "Lise." Sansom's initial objective was to contact the French Resistance on the French Riviera, and then move to Auxerre in Burgundy to establish a safe house for other agents.

Adolphe Rabinovitch

At the time of her arrival in France, the Spindle network was beset by internal strife between the principal agent, André Girard, and his assistant and with the network's radio operator, Adolphe Rabinovitch. A list of 200 potential supporters, lost by André Marsac, a Girard courier, was obtained by the Germans. With Sansom stranded in

We Live in a Malleable Reality-and We Can Change It

Cannes, Churchill obtained Buckmaster's permission to scrap her original mission and for her to act as his courier.

Sansom, posing as "Madame Odette Metayer," was required to find food and lodging for Rabinovitch, who was in France illegally and had no ration card, and also to tend to air drops that were sometimes carelessly placed in dangerous areas. Her work brought her initially to Marseilles, then considered a dangerous town because of its infiltration by German agents. Sansom was shocked by the lax attitude toward security by her French supporters. Sansom grew close to Churchill and to Rabinovitch, whom she liked and trusted. She later recalled that she had suspicions of disloyalty about other members of the Spindle network, but declined to identify who she suspected.

Captured

In January 1943, the Spindle team of Churchill, Rabinovich, and Sansom, feeling vulnerable to German capture, moved north from the French Riviera to the quiet Italian-occupied Annecy area in the French Alps. Churchill and Sansom took up residence at the Hotel de la Poste in the village of Saint-Jorioz. They were joined there by several other members of the Carte network and SOE, a gathering which attracted the attention of the Italian fascist police and the Gestapo. SOE agent Francis Cammaerts visited Annecy briefly in March or early April 1943 and assessed the security of Churchill and Sansom's network as deficient and likely to be penetrated by the Germans.

Meanwhile, in Paris in mid-March, spy-catcher Hugo Bleicher, an Abwehr counterintelligence officer, arrested Marsac, persuading him and another Carte associate, Roger Bardet, that he was an anti-Nazi German colonel

We Live in a Malleable Reality-and We Can Change It

and that they should work together. He learned from Marsac the location of Churchill and Sansom, got a letter of introduction to them from him, and proceeded to Saint-Jorioz where he introduced himself to Sansom as "Colonel Henri." He spun a tale to her of them traveling together to London to "discuss means of ending the war." He then departed Saint-Jorioz with a plan to return and depart France together by clandestine aircraft on April 18. Sansom had Rabinovich send a wireless message to SOE headquarters in London reporting the contact. London replied immediately: "Henri highly dangerous...you are to hide across lake and cut contacts with all save Arnaud [Rabinovich]..."

Churchill was in London consulting with SOE at the time of Bleicher's meeting with Sansom. He was warned to avoid contact with Sansom and 'Colonel Henri" on his return to France, but when he was parachuted back into the Annecy area on April 14/15, he was met by Sansom and Rabinovich. As Sansom did not anticipate the return of Henri until April 18, she and Churchill proceeded to the hotel in Saint-Jorioz. At 2:00 a.m. on April 16, Bleicher, no longer in the guise of "Colonel Henri," appeared in the hotel with Italian soldiers and arrested Sansom and Churchill.

Imprisonment-Fresnes Prison

At Fresnes prison, near Paris, Sansom was interrogated by the Gestapo fourteen times. She was subjected to torture. Her back was scorched with a red-hot poker and all of her toenails were pulled out. She refused to disclose the whereabouts of Rabinovitch and another British agent, stuck to her fabricated cover story that Churchill was the nephew of Prime Minister Winston Churchill, that she was his wife, and that he knew nothing of her activities. The

We Live in a Malleable Reality-and We Can Change It

hope was that in this way their treatment would be mitigated. The British had calculated that if the Germans thought she was related to the British Prime Minister, they would want to keep her alive as a possible bargaining tool.

Sansom succeeded in diverting attention from Churchill, who was subject to only two interrogations, and protected the identities of the two officers, whose locations were known only to her. While imprisoned, Bleicher occasionally appeared and sought to invite her to travel with him to Paris to attend concerts and dine in restaurants, to persuade her to talk. Sansom rejected the overtures.

She was condemned to death on two counts in June 1943, to which she responded, "Then you will have to make up your mind on what count I am to be executed, because I can only die once." Infuriated, Bleicher sent her to Ravensbrück concentration camp.

Ravensbruck concentration camp

In Ravensbruck Sansom was kept in a punishment block cell, on a starvation diet, and could hear other prisoners being beaten.

After the Allied landings in the south of France in August 1944, on orders from Berlin, all food was withdrawn for a week, all light was removed from Sansom's cell, and the heat was turned up. Despite a report by the camp doctor that she would not survive such conditions for more than a few weeks, after being found unconscious in her cell she was placed in solitary confinement. Her conditions only improved in December 1944, when she was moved to a ground floor cell. The cell was located near the crematorium and would be covered with burned hair from the cremations. At one point toward the end of the war,

We Live in a Malleable Reality-and We Can Change It

she witnessed an instance of cannibalism of a dead inmate by starving prisoners.

When the Allies were only a few miles from Ravensbrück, the camp commandant Fritz Suhren took Sansom and drove to an American base to surrender. He hoped that her supposed connections to Churchill might allow him to negotiate his way out of execution.

Churchill survived the war but Rabinovitch was executed by the Gestapo in 1944.

Survival methods

Sansom was aided in her endurance in prison by her early blindness and paralysis, and by the example of her grandfather, who "did not accept weakness very easily." She also accepted in advance that she might be captured by the Germans. She adopted an attitude of defiance, and found that this resulted in a degree of respect by her captors and helped her survive the imprisonment mentally.

Sansom said she believed she was "not brave, not courageous, but just make up my mind about certain things." She recalled in a post-war interview that while everyone has a breaking point, her feeling was that if she could "survive the next minute without breaking up, that is another minute of life. And if I can think that way instead of thinking what is going to happen in a half-hour's time," Because of her past illnesses she knew "I was able to accept this, and survive it." By accepting death, she felt that "they would not win anything. They'll have a dead body, useless to them. They won't have me. I won't let them have me." She described it as a "kind of bargaining." The Germans generally found persons of the prisoners' own nationality to carry out the torture, she later recalled,

We Live in a Malleable Reality-and We Can Change It

so that one "could not say they were tortured by the Germans." Her torture was carried out by a "very good-looking young Frenchman" who she believed was mentally ill.

We Live in a Malleable Reality-and We Can Change It

11.3 Ernest Shakelton's Voyage to the South Pole

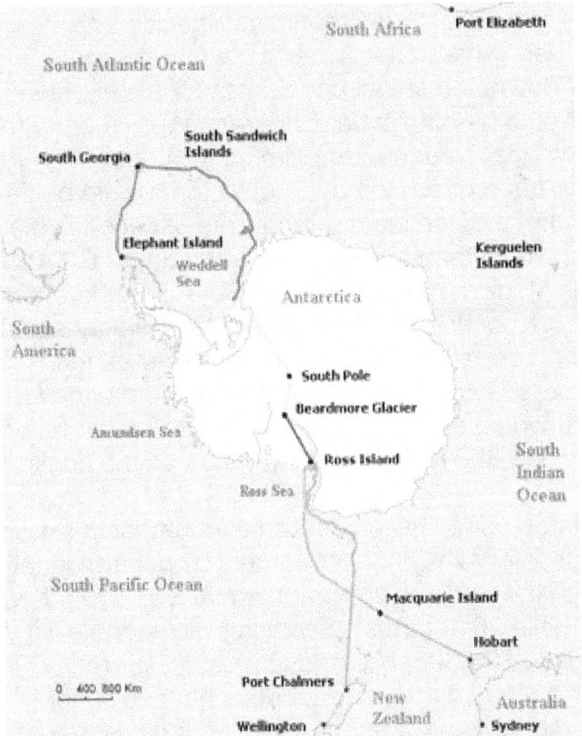

Imperial Trans-Antarctic Expedition, 1914–1917

Preparations

Main articles: Imperial Trans-Antarctic Expedition, Ross Sea Party, and List of personnel of the Imperial Trans-Antarctic Expedition

Outline of Antarctica coast, with different lines indicating the various journeys made by ships and land parties during the expedition

We Live in a Malleable Reality-and We Can Change It

Shackleton published details of his new expedition, grandly titled the "Imperial Trans-Antarctic Expedition", early in 1914. There is a legend that says Shackleton's newspaper article was written a certain way so that he could better narrow down and select candidates for his expedition. Two ships would be employed; Endurance would carry the main party into the Weddell Sea, aiming for Vahsel Bay from where a team of six, led by Shackleton, would begin the crossing of the continent. Meanwhile, a second ship, the Aurora, would take a supporting party under Captain Aeneas Mackintosh to McMurdo Sound on the opposite side of the continent. This party would then lay supply depots across the Great Ice Barrier as far as the Beardmore Glacier; these depots would hold the food and fuel that would enable Shackleton's party to complete their journey of 1,800 miles (2,900 km) across the continent.

Shackleton used his considerable fund-raising skills, and the expedition was financed largely by private donations, although the British government gave £10,000 (about £900,000 in 2019 terms). Scottish jute magnate Sir James Caird gave £24,000, Midlands industrialist Frank Dudley Docker gave £10,000, and tobacco heiress Janet Stancomb-Wills gave an undisclosed but reportedly "generous" sum. Public interest in the expedition was considerable; Shackleton received more than 5,000 applications to join it.

His interviewing and selection methods sometimes seemed eccentric; believing that character and temperament were as important as technical ability, he asked unconventional questions. Thus physicist Reginald James was asked if he could sing; others were accepted on sight because Shackleton liked the look of them, or after the briefest of interrogations. Shackleton also loosened some traditional hierarchies to promote

We Live in a Malleable Reality-and We Can Change It

camaraderie, such as distributing the ship's chores equally among officers, scientists, and seamen. He also socialized with his crew members every evening after dinner, leading sing-alongs, jokes, and games. He ultimately selected a crew of 56, twenty-eight on each ship.

Despite the outbreak of the First World War on 3 August 1914, Endurance was directed by the First Lord of the Admiralty, Winston Churchill, to "proceed", and left British waters on 8 August. Shackleton delayed his own departure until 27 September, meeting the ship in Buenos Aires.

Crew

While Shackleton led the expedition, Captain F. Worsley commanded the Endurance and Lieutenant J. Stenhouse the Aurora. On the Endurance, the second in command was the experienced explorer Frank Wild. The meteorologist was Captain L. Hussey, also an able banjo player. McIlroy was head of the scientific staff, which included Wordie.

Alexander Macklin was one of two surgeons and also in charge of keeping the 70 dogs healthy. Tom Crean was in more immediate charge as head dog-handler. Other crew included James, Hussey, Greenstreet, a carpenter Henry McNeish, and a biologist named Clark. Of later independent fame was the photographer Frank Hurley, known on this mission for his perilous shots.

There was a (male) cat named Mrs Chippy that belonged to the carpenter Henry McNeish. Mrs Chippy was shot when the Endurance sank, due to the belief that he would not have survived the ordeal that followed.

We Live in a Malleable Reality-and We Can Change It

Loss of Endurance

Endurance departed from South Georgia for the Weddell Sea on 5 December, heading for Vahsel Bay. As the ship moved southward navigating in ice, first year ice was encountered, which slowed progress. Deep in the Weddell Sea, conditions gradually grew worse until, on 19 January 1915, Endurance became frozen fast in an ice floe.

On 24 February, realising that she would be trapped until the following spring, Shackleton ordered the abandonment of ship's routine and her conversion to a winter station. She drifted slowly northward with the ice through the following months. When spring arrived in September, the breaking of the ice and its later movements put extreme pressures on the ship's hull.

Shackleton after the loss of Endurance

Until this point, Shackleton had hoped that the ship, when released from the ice, could work her way back towards Vahsel Bay. On 24 October, water began pouring in. After a few days, with the position at 69° 5' S, 51° 30' W, Shackleton gave the order to abandon ship, saying, "She's going down!"; and men, provisions and equipment were transferred to camps on the ice. On 21 November 1915, the wreck finally slipped beneath the surface.

For almost two months, Shackleton and his party camped on a large, flat floe, hoping that it would drift towards Paulet Island, approximately 250 miles (402 km) away, where it was known that stores were cached. After failed attempts to march across the ice to this island, Shackleton decided to set up another more permanent camp (Patience Camp) on another floe, and trust to the drift of the ice to take them towards a safe landing.

We Live in a Malleable Reality-and We Can Change It

By 17 March, their ice camp was within 60 miles (97 km) of Paulet Island; however, separated by impassable ice, they were unable to reach it. On 9 April, their ice floe broke into two, and Shackleton ordered the crew into the lifeboats and to head for the nearest land.

After five harrowing days at sea, the exhausted men landed their three lifeboats at Elephant Island, 346 miles (557 km) from where the Endurance sank. This was the first time they had stood on solid ground for 497 days. Shackleton's concern for his men was such that he gave his mittens to photographer Frank Hurley, who had lost his during the boat journey. Shackleton suffered frostbitten fingers as a result.

Open-boat journey

Launching the James Caird from the shore of Elephant Island, 24 April 1916 Elephant Island was an inhospitable place, far from any shipping routes; rescue by means of chance discovery was very unlikely. Consequently, Shackleton decided to risk an open-boat journey to the 720-nautical-mile-distant South Georgia whaling stations, where he knew help was available. The strongest of the tiny 20-foot (6.1 m) lifeboats, christened James Caird after the expedition's chief sponsor, was chosen for the trip. Ship's carpenter Harry McNish made various improvements, including raising the sides, strengthening the keel, building a makeshift deck of wood and canvas, and sealing the work with oil paint and seal blood.

Shackleton chose five companions for the journey: Frank Worsley, Endurance's captain, who would be responsible for navigation; Tom Crean, who had "begged to go"; two strong sailors in John Vincent and Timothy McCarthy, and finally the carpenter McNish. Shackleton had clashed with

We Live in a Malleable Reality-and We Can Change It

McNish during the time when the party was stranded on the ice, but, while he did not forgive the carpenter's earlier insubordination, Shackleton recognized his value for this particular job. Not only did Shackleton recognize their value for the job but also because he knew the potential risk they were to morale. This allowed for Shackleton to remain in control of the morale of his crew members. The attitudes of his men were a point of emphasis in leading his men back to safety.

Shackleton refused to pack supplies for more than four weeks, knowing that if they did not reach South Georgia within that time, the boat and its crew would be lost. The James Caird was launched on 24 April 1916; during the next fifteen days, it sailed through the waters of the southern ocean, at the mercy of the stormy seas, in constant peril of capsizing. On 8 May, thanks to Worsley's navigational skills, the cliffs of South Georgia came into sight, but hurricane-force winds prevented the possibility of landing. The party was forced to ride out the storm offshore, in constant danger of being dashed against the rocks. They later learned that the same hurricane had sunk a 500-ton steamer bound for South Georgia from Buenos Aires.

On the following day, they were able, finally, to land on the unoccupied southern shore. After a period of rest and recuperation, rather than risk putting to sea again to reach the whaling stations on the northern coast, Shackleton decided to attempt a land crossing of the island. Although it is likely that Norwegian whalers had previously crossed at other points on ski, no one had attempted this particular route before. For their journey, the survivors were only equipped with boots they had pushed screws into to act as climbing boots, a carpenter's adze, and 50 feet of rope. Leaving McNish, Vincent and McCarthy at the landing

We Live in a Malleable Reality-and We Can Change It

point on South Georgia, Shackleton travelled 32 miles (51 km) with Worsley and Crean over extremely dangerous mountainous terrain for 36 hours to reach the whaling station at Stromness on 20 May.

The next successful crossing of South Georgia was in October 1955, by the British explorer Duncan Carse, who travelled much of the same route as Shackleton's party. In tribute to their achievement, he wrote: "I do not know how they did it, except that they had to — three men of the heroic age of Antarctic exploration with 50 feet of rope between them – and a carpenter's adze".

Rescue

Shackleton immediately sent a boat to pick up the three men from the other side of South Georgia while he set to work to organize the rescue of the Elephant Island men. His first three attempts were foiled by sea ice, which blocked the approaches to the island. He appealed to the Chilean government, which offered the use of the Yelcho, a small seagoing tug from its navy. Yelcho, commanded by Captain Luis Pardo, and the British whaler Southern Sky reached Elephant Island on 30 August 1916, at which point the men had been isolated there for four and a half months, and Shackleton quickly evacuated all 22 men. The Yelcho took the crew first to Punta Arenas and after some days to Valparaiso in Chile where crowds warmly welcomed them back to civilization.

There remained the men of the Ross Sea Party, who were stranded at Cape Evans in McMurdo Sound, after Aurora had been blown from its anchorage and driven out to sea, unable to return. The ship, after a drift of many months, had returned to New Zealand. Shackleton travelled there to join Aurora, and sailed with her to the rescue of the

We Live in a Malleable Reality-and We Can Change It

Ross Sea party. This group, despite many hardships, had carried out its depot-laying mission to the full, but three lives had been lost, including that of its commander, Aeneas Mackintosh.

We Live in a Malleable Reality-and We Can Change It

12.0   Spiritual and Energy Healing

There are many aspects of Spiritual and Energy Healing as described in my book "The Handbook of Spiritual and Energy Healing: And How to Learn it Yourself". Here is just one technique which you can use to practice healing visualizations:

Healing visualizations

The holistic practice of creating positive images of healing within the mental body, which allows the emotional body to respond by becoming receptive to the healing, which subsequently initiates the healing process in the physical body.

Our body and mind are intricately connected and our thoughts have a direct influence on our physical health. By using the power of our mind to envision our healing process, we're disciplining our nervous system to react in a positive way, resulting in the desired outcome.

The regular practice of healing visualizations allows us to detach from our past limited beliefs to release our fears

We Live in a Malleable Reality-and We Can Change It

and bring our awareness to the present moment to enable us to create a healthy future. Traditionally, there was an innate awareness of the mind, body and soul connection and healing visualizations were practiced through daily prayer.

The Empowerment Healing Visualization relieves stress, anxiety, fear and panic.

Practice this two times per day.

- Sit outdoors in nature, with your feet in direct contact with the bare earth, grass or sand.
- Close your eyes and visualize the colour green.
- See the colour green as a dense vortex of deep green healing energy.
- Sense the dense vortex of deep green healing energy spiraling in the earth, beneath your feet.
- Feel its magnetic force gently drawing your energies downwards, through the soles of your feet.
- Visualize yourself being rooted into the earth below.
- Allow your entire being to feel at one with nature.
- Bring your awareness to your body.
- Your body feels safe and grounded by the deep green healing energy.
- Bring your awareness to your mind.
- Your mind feels still and centered by the deep green healing energy.
- Bring your awareness to your emotions.
- Your emotions feel balanced and peaceful by the deep green healing energy.
- Bring your awareness to your breath.
- Visualize the dense vortex of deep green healing energy, attracting your inhalations, pulling them deeper within your body.

We Live in a Malleable Reality-and We Can Change It

- As your breathing becomes deeper, your body becomes relaxed, your mind becomes calm and your emotions become balanced.
- You feel grounded, centered and in control.
- Observe this feeling of empowerment within your body, mind and emotions.
- Create a positive image of this feeling of empowerment within your body, your mind and your emotions.
- Draw upon it during times of stress, anxiety, fear and panic.
- Open your eyes gently and repeat aloud,
  "I trust in the power of my breath to instill strength, stability and harmony throughout my entire being."

As we strive to meet the demands of our busy lifestyles, we can easily become detached from nature, yet our inner wisdom intuitively reminds us to care for our body and mind holistically. By re-connecting with the ancient Indian wisdom of self-healing practices, we can cleanse, calm and empower our energies naturally, to restore balance, harmony and peace of mind.

We Live in a Malleable Reality-and We Can Change It

We Live in a Malleable Reality-and We Can Change It

13.0 Spiritual Development and Psychic Abilities

In my book "God Like Powers and Abilities" I went into detail about how the spiritual development process has the side effect of students developing psychic and paranormal abilities. Here are more details about how and why this works:

13.1 The Yoga Sutras of Patanjali

I first learned about the read the Yoga Sutras when I was about nineteen, and it has impressed me to this day as a well written exposition for enlightenment and an analytical view of Yoga.

I believe that the Sutras also describe the underlying reality behind all religions and mystical philosophies.

"The Yoga Sutras of Patanjali" was written in four books in the Sanskrit language and is one of the great spiritual works of India.

It is available in print and in many versions on the Internet, and describes the path to enlightenment and attainment of spiritual powers.

We Live in a Malleable Reality-and We Can Change It

The Sutras were written over 2,000 years ago by some estimates.

The Sutras are often described as a scientific exposition in the science of yoga and consists of 4 books of Sutras in Sanskrit which has numerous translations in English.

Each book was written in short sentences of Sanskrit, so many publications of the Yoga Sutras today include the following:

1) The original Sanskrit text
2) The phonetic and exact word translation
3) Turning the previous into and English sentence
4) An exposition on the meaning and details of the Sutra.

AN EXAMPLE OF AN ANCIENT SANSKRIT DOCUMENT

All of the superhuman abilities and powers and more referred to in the Sutras are found in book three.

The goal of the yogic path and the Sutras is enlightenment, however many spiritual powers become side effects of this spiritual process.

Many Sutras describe different abilities or how different abilities can be derived.

We Live in a Malleable Reality-and We Can Change It

As an example-although the Sutras don't talk specifically of physical immortality, the powers available as a result of the spiritual development process would also affect the body in a very positive manner.

In Book 3 Sutra 45 we have an example of what I'm referring to regarding spiritual powers from the process of enlightenment which would probably cause a lot of longevity in the physical body if exercised properly:

Book 3 Sutra 45-Thereupon will come the manifestation of the atomic and other powers, which are the endowment of the body, together with its unassailable force.

The body in question is, of course, the etheric body of the spiritual man.

He is said to possess eight powers: the atomic, the power of assimilating himself with the nature of the atom, which will, perhaps, involve the power to disintegrate material forms; the power of levitation; the power of limitless extension; the power of boundless reach, so that, as the commentator says, "he can touch the moon with the tip of his finger"; the power to accomplish his will; the power of gravitation, the correlative of levitation; the power of command; the power of creative will.

These are the endowments of the spiritual man. Further, the spiritual body is unassailable. Fire burns it not, water wets it not, the sword cleaves it not, dry winds parch it not.

We Live in a Malleable Reality-and We Can Change It

And, it is said, the spiritual man can impart something of this quality and temper to his bodily vesture.

Many forms of Yoga exist to develop one along the path. These include Raja, Gnana, Praya, Tantric, and additional forms of Yoga.

Yoga teachers for many of these disciplines can be found in many major and minor cities around the world.

We Live in a Malleable Reality-and We Can Change It

13.2 How Superhuman Abilities work

To understand superhuman abilities you need to have a metaphysical or Eastern philosophical understanding of the universe.

The belief and experiences of adepts in the East is that the reality that we experience all of the time is really an illusion.

Underlying this illusion is the life force or "God" of the universe which exists outside of time and space.
In fact, the basic tenet of Eastern Philosophy is that the spirit at the core of our being is part of the same underlying living spirit aspect of the Universe.

Even Christianity can be interpreted to say that God's spirit within us is really that God force at the core of our being which is also outside of time and space.

In the Chapter on Prophecy we will get into more details about what it means to be outside of time and space and why advanced adepts work to get in close touch with their spirits.

We Live in a Malleable Reality-and We Can Change It

This is because the illusion of the world or "samsara" can be penetrated when you live more in the spirit as part of the process of enlightenment.

As one understands at a deeper and deeper level the underlying spiritual reality of stillness, the more one can manipulate the "reality illusion" to obtain the abilities of the spirit. It's very similar to the concept underlying the Matrix movies.

When you fully understand the underlying reality of spirit, you can do things in what most people call "reality" which would seem impossible without that deeper understanding. Another analogy is the ability to create an atomic weapon. The results of blowing up atomic weapons would seem as the power of God to a person ignorant of Nuclear Physics. One you understand the physics, then it's just an engineering challenge to build the weapon.

Similarly, once your spirit understands the underlying reality of spirit, then it's just learning the correct techniques and processes to acquire what seem to be superhuman abilities.

We Live in a Malleable Reality-and We Can Change It

2) An understanding of and the ability to manipulate the vital force or prana which is in all of our energy bodies.

A Meditation Exercise

Meditation is best learned from a teacher since that process provides regular feedback to help you improve your technique.

However, for those of you who don't have immediate access to a Teacher, here are some basic steps to starting your own meditation program:

Step One-Set a regular time daily of 30 mins to do your mediation. Some persons like to do it early after they get up, or at a free time later in the day. Should not be just before bedtime because you may fall asleep and not get full benefit from the meditation session.

Step Two-Make yourself comfortable. It can be sitting or lying down. You don't have to do a cross legged sitting position unless you want to.

Step Three-Start by closing your eyes and working to relax your body.

Step Four-This can be done by going to each portion of your body and telling it to tighten then relax. An example would be: tighten you hand..tight..tight.. then relax. Do this with each portion of your body.

Step Five-After your body is fully relaxed then you need to relax your mind.

Step Six-You can do this by visualizations such as—my thoughts are slowing down, my thoughts are being

We Live in a Malleable Reality-and We Can Change It

13.3 The Path to Supernormal Abilities

As I've discussed in my other books, it's my firm belief based on study and personal experience that the path to all spiritual powers is through learning stillness.
Learning stillness is accomplished through meditation and deep prayer; and is the path to spiritual development.
It is mainly persons who are starting to understand their true nature who begin to exhibit amazing abilities.
The danger is to not get sidetracked on using these powers and abilities since they can entrance one into focusing on these activities and on building the Ego.

Unraveling the Ego is the goal of stillness practices, while rebuilding the Ego will ruin the spiritual development of the seeker and set them back many years.

Learning superhuman abilities can also be an aid to spiritual development if used properly.

The ability to perform these superhuman skills can solidify your faith and belief in the spirit as being at the foundation of reality.

I think the reason we don't see many of these powers and abilities demonstrated in our everyday lives is that they are mostly available to the spiritually adept who know the harm in using these abilities except when necessary.
To successfully accomplish most of the paranormal ability exercises in this book you will need two things:

1) An understanding and practical experience with meditation or deep prayer. These practices involve how to relax your body and focus your attention on the exercises.

We Live in a Malleable Reality-and We Can Change It

13.2 How Superhuman Abilities work

To understand superhuman abilities you need to have a metaphysical or Eastern philosophical understanding of the universe.

The belief and experiences of adepts in the East is that the reality that we experience all of the time is really an illusion.

Underlying this illusion is the life force or "God" of the universe which exists outside of time and space.
In fact, the basic tenet of Eastern Philosophy is that the spirit at the core of our being is part of the same underlying living spirit aspect of the Universe.

Even Christianity can be interpreted to say that God's spirit within us is really that God force at the core of our being which is also outside of time and space.

In the Chapter on Prophecy we will get into more details about what it means to be outside of time and space and why advanced adepts work to get in close touch with their spirits.

We Live in a Malleable Reality-and We Can Change It

This is because the illusion of the world or "samsara" can be penetrated when you live more in the spirit as part of the process of enlightenment.

As one understands at a deeper and deeper level the underlying spiritual reality of stillness, the more one can manipulate the "reality illusion" to obtain the abilities of the spirit. It's very similar to the concept underlying the Matrix movies.

When you fully understand the underlying reality of spirit, you can do things in what most people call "reality" which would seem impossible without that deeper understanding. Another analogy is the ability to create an atomic weapon. The results of blowing up atomic weapons would seem as the power of God to a person ignorant of Nuclear Physics. One you understand the physics, then it's just an engineering challenge to build the weapon.

Similarly, once your spirit understands the underlying reality of spirit, then it's just learning the correct techniques and processes to acquire what seem to be superhuman abilities.

We Live in a Malleable Reality-and We Can Change It

released as little balloons. I'm sinking and relaxing like sinking down in peaceful warm water.

Step Seven-When you get to a very relaxed level you should imagine your Spirit is in no space and no time—to start working towards a level of Stillness. Another way of visualizing this is that you are existing only in the present—no past and no future.

Step Eight-After practicing this for a couple of weeks you should be able to start relaxing immediately to a fairly deep state.

Again, there are many books on how to meditate and many instructors.
This practice takes years to really get to a deep state, but starting now will quickly begin to make a difference in your body's spiritual and physical health.

We Live in a Malleable Reality-and We Can Change It

We Live in a Malleable Reality-and We Can Change It

released as little balloons. I'm sinking and relaxing like sinking down in peaceful warm water.

Step Seven-When you get to a very relaxed level you should imagine your Spirit is in no space and no time—to start working towards a level of Stillness. Another way of visualizing this is that you are existing only in the present—no past and no future.

Step Eight-After practicing this for a couple of weeks you should be able to start relaxing immediately to a fairly deep state.

Again, there are many books on how to meditate and many instructors.
This practice takes years to really get to a deep state, but starting now will quickly begin to make a difference in your body's spiritual and physical health.

We Live in a Malleable Reality-and We Can Change It

We Live in a Malleable Reality-and We Can Change It

14.0 Summary

In my late teens and early twenties I learned a lot about spiritual growth, enlightenment, and paranormal abilities. Between this knowledge and what I learned in engineering school along with lots of physics I thought I had a pretty good understanding our reality.

In recent years however, as I continue to research unusual topics like time travel, parallel dimensions, and more, I feel more and more ignorant of the true reality we live in.

Now, I lean much more to the idea that there are many things we don't know about our universe and the interfaces it has with other parts of reality.

We can manipulate our reality and universe as illustrated by the samples of many experiences I've provided in this book.

There are real limits on what we can learn about the universe we live in, but we are nowhere close to these limits. In fact our spiritual abilities may be able to help us better understand the parts of this universe which are hard to measure otherwise.

Our reality is not as fixed and understood as scientists and engineers would have us believe.

Therefore, we need to be open minded but logical about what does exist in our world no matter how strange it seems. This open-mindedness will lead us to new discoveries to better understand the world we live in.

We Live in a Malleable Reality-and We Can Change It

All the Best,

Martin K. Ettington
September 2021

We Live in a Malleable Reality-and We Can Change It

15.0 Bibliography

1. https://www.science.org/news/2017/10/quantum-experiment-space-confirms-reality-what-you-make-it-0. *Quantum Experiment Confirms Reality is what you Make it.* [Online]

2. **Ettington, Martin K.** *Our Energy Bodies, Auras, and Thoughtforms.*

3. —. *Physical Immortality: A History and How to Guide.*

4. —. *Real Time Travel Stories from a Psychic Engineer.*

5. —. *Stories of Parallel Dimensions.*

6. —. *On Using the Scienctific Method to Study the Paranormal.*

7. —. *Use Intuition and Prophecy to Improve Your Life By An Adept.*

8. —. *The Handbook of Spiritual and Energy Healing: And How to Learn it Yourself.*

9. —. *Building and Stocking a Nuclear Shelter for Under $10,000.*

10. —. *The God Like Powers and Abilities.*